The Facts on Matter

What are MIXTURES?

By Elise Tobler

Please visit our website, www.garethstevens.com. For a free color catalog of all our high-quality books, call toll free 1-800-542-2595 or fax 1-877-542-2596.

Library of Congress Cataloging-in-Publication Data

Names: Tobler, Elise, 1970- author.
Title: What are mixtures? / Elise Tobler.
Description: New York : Gareth Stevens Publishing, [2022] | Series: The facts on matter | Includes index.
Identifiers: LCCN 2020031192 (print) | LCCN 2020031193 (ebook) | ISBN 9781538267097 (library binding) | ISBN 9781538267073 (paperback) | ISBN 9781538267080 (set) | ISBN 9781538267103 (ebook)
Subjects: LCSH: Solution (Chemistry)–Juvenile literature. | Mixtures–Juvenile literature.
Classification: LCC QD541 .T62 2022 (print) | LCC QD541 (ebook) | DDC 541/.34–dc23
LC record available at https://lccn.loc.gov/2020031192
LC ebook record available at https://lccn.loc.gov/2020031193

First Edition

Published in 2022 by
Gareth Stevens Publishing
29 E. 21st Street
New York, NY 10010

Designer: Katelyn E. Reynolds
Editor: Char Light

Photo credits: Cover, pp. 1 (mixture), 13 ttsz/ iStock / Getty Images Plus; cover, pp. 1–24 (background) sudanmas/E+/Getty Images; cover, pp. 1–24 (chemistry doodles) backUp/Shutterstock.com; cover, pp. 1–24 (banner) Ozz Design/Shutterstock.com; cover, pp. 1–24 (paper) Sergey Mironov /Shutterstock.com; p. 5 fcafotodigital/ iStock / Getty Images Plus; p. 7 Simon McGill/ Moment/Getty Images; p. 9 (top) John E. Kelly/ Photolibrary / Getty Images Plus; p. 9 (bottom) MarkSwallow/E+/Getty Images; p. 10 Dirk Meister/ Moment/Getty Images; p. 13 Burke/Triolo Productions/ Photolibrary / Getty Images Plus; p. 15 (top) dszc/ iStock / Getty Images Plus; p. 15 (bottom) imran kadir photography/ Moment/Getty Images; p. 17 (top) guruXOX/ Shutterstock.com; p. 17 (bottom) amixstudio/ iStock / Getty Images Plus; p. 19 Pat_Hastings/Shutterstock.com; p. 21 inewsfoto/ Shutterstock.com.

Printed in the United States of America

CPSIA compliance information: Batch #CWGS22: For further information contact Gareth Stevens, New York, New York at 1-800-542-2595.

CONTENTS

Words in the glossary appear in **bold** type the first time they are used in the text.

What Is a MIXTURE?

In **chemistry**, a mixture is a **physical** combination of two or more things. A **chemical reaction** doesn't need to happen to make a mixture. If you have pink candies and blue candies, and you put them in one bowl, you've made a mixture! Mixtures don't need to have equal parts—you can have more blue candies than pink candies in your bowl.

The parts that make up a mixture keep their own **properties**. This means mixtures can be separated physically.

You could take this mixture of candies apart by sorting each type of candy into groups.

Solid MIXTURES

Mixtures can happen in any state of matter. Solid mixtures are often combinations of compounds. A compound is made of two or more elements that have bonded, or chemically joined.

Concrete is made of water (a compound of the elements hydrogen and oxygen), rocks or sand, and cement (a paste that hardens). These parts combine and grow hard to form a solid mixture. You can see the different kinds of rock inside a concrete sidewalk. If we break the concrete, each rock can be separated.

Know the Basics!

Matter comes in four states: solid, liquid, gas, and **plasma**. Each state has different properties. For example, ice cubes can hold their own shape (while they're cold enough), but liquid water cannot.

Concrete is a mixture and not a compound. Its parts are held together physically. Compounds are held together chemically.

Liquid MIXTURES

Liquids can't hold their own shape. If they are in a container, they take on the shape of that container. Your blood is one example of a liquid mixture. The parts that mix to make your blood can be separated.

An example of a liquid and solid mixture is ice cubes in soda. When mixed, you can see each part separately. You can easily take the ice cubes and soda apart and put them back together.

Know the Basics!

Polluted water is a mixture, combining water and pollution. We have the ability to clean the pollution out of the water by **filtering** it, leaving the water clean and drinkable.

Your breakfast cereal in milk is a solid and liquid mixture.

Gas MIXTURES

The air you're breathing right now is an example of a gas mixture. Air is made up of the elements oxygen, nitrogen, carbon dioxide, and other gases. Each gas can be separated out.

Gases always mix with other gases homogeneously, or evenly. This means the particles, or tiny pieces, of each gas are evenly spread throughout the mixture. At the same time, the lightest gases will be at the "top" of the mixture, while the heaviest gases will be at the "bottom" of the mixture.

Air pollution is a mixture of gases combining to create brown clouds across cities.

EXAMPLES OF MIXTURES BY STATES OF MATTER

	liquid	solid	gas
liquid	blood	ice cubes in a glass of water	fog, a mixture of air and water
solid	mud, a mixture of dirt and water	a bowl of different fruit pieces	smoke, a mixture of tiny pieces of soot (powder that comes from burning items) and air
gas	carbon dioxide in soda	palladium hydride, a mixture of the metal palladium and the gas hydrogen	gases inside a neon light

Fog and smoke are both aerosols, or "aero-solutions." Aerosols are mixtures made of tiny pieces of a liquid or solid floating in gas.

Homogeneous MIXTURES

A homogeneous mixture contains evenly spread parts. If each item within the mixture cannot be seen well once it's mixed, the mixture is homogeneous. You cannot see two states of matter in a homogeneous mixture.

Many homogeneous mixtures are liquids. These are also called solutions. Saltwater and gasoline are examples of liquid homogeneous mixtures, or solutions. Some solids are also homogeneous mixtures. Plastic and wood are solid homogeneous mixtures, each part smoothly blended. Even the gas in neon lights is often a homogeneous mixture.

Know the Basics!

Suspensions are a type of mixture. They are liquids with solids floating in them. Oil and water is probably the best example. Drops of oil are **suspended** in the water until they are shaken and blended together.

OIL + WATER

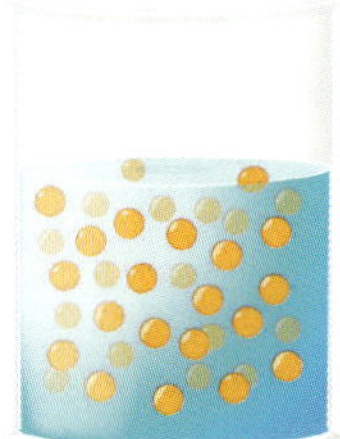

"Homogenized milk" means the fat molecules that would make cream are broken up and spread evenly through the liquid.

Heterogeneous MIXTURES

Heterogeneous mixtures have their **components** unevenly spread throughout them. There are always two or more states of matter—like solid ice cubes in a liquid, or solid cereal in liquid milk.

Chocolate chip cookie dough is another example of a heterogeneous mixture. You can easily see the chocolate in the dough. The individual **ingredients** in a heterogeneous mixture do not have to be evenly added. When you bite into a cookie, you may get a different number of chips every time.

Cookie dough by itself is a homogeneous mixture. The addition of nuts or chocolate chips makes it a heterogeneous mixture.

Alloys are another kind of mixture. They can be a homogeneous or heterogeneous mixture.

Filtration and CENTRIFUGES

If you have a mixture of sand and water, you can pour the water through a filter, leaving only the sand behind. This is called filtration. We filter liquids to remove unsafe particles, or small bits.

You can also separate mixtures with a machine called a centrifuge. Centrifuges spin fast, which makes liquids separate. Centrifuges are used to separate the liquid parts of blood from the solid parts. You can also use a centrifuge to separate cream from milk.

Know the Basics!

When a centrifuge spins, the lighter part of a mixture rises to the top of the container, and the heavier part drops to the bottom.

Different types of centrifuges can separate gases, metals, and even the water from your clean laundry.

Evaporation and DISTILLATION

Evaporation is when a liquid turns to a vapor, or gas, without **boiling**. You can see evaporation by watching paint dry. Paint is liquid when you brush it onto paper. As the water evaporates, it becomes more of a solid.

Another way to separate mixtures is through a process called distillation. When two liquids have different temperatures at which they boil, they can be separated. One part of the mixture will turn to vapor, leaving the other part of the mixture behind.

Know the Basics!

Magnetism is another way to separate some mixtures. Magnets are metals that can pull other metals toward them. If you mix metal with sand, you can easily pull the metal out with a magnet.

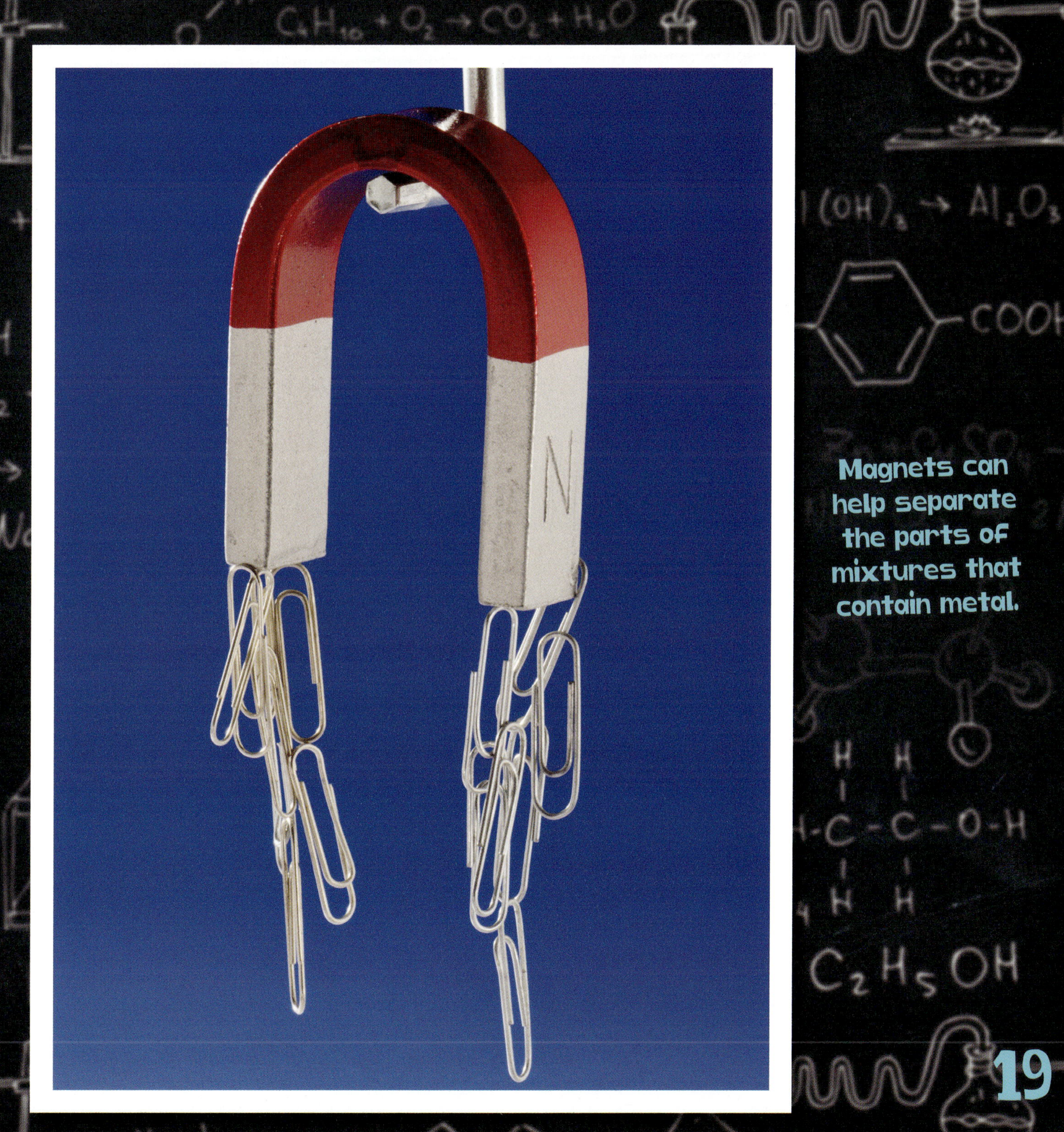

Magnets can help separate the parts of mixtures that contain metal.

Food MIXTURES

Milk is a mixture called a colloid. A colloid is one substance spread inside another, where the suspended particles never sink to the bottom. For milk, this means it contains tiny pieces of floating cream. Mayonnaise is another example of a colloid—it's tiny bits of egg yolk floating in oil!

Colloids can be physically separated. When you shake cream in a jar, the floating fats will separate from the liquid. Then the fats join together and turn into butter!

Some mixtures don't need fancy tools to be separated. You can often pick them apart with your fingers. Salad is a mixture—if you don't like one of the toppings, you can pick it right off!

GLOSSARY

alloy: a metal made by melting and mixing two or more metals or a metal and another material together

boil: to heat a liquid until it forms bubbles

chemical reaction: when a set of substances changes to make a new substance

chemistry: a science that deals with matter and the changes it goes through

component: one of the parts of something

filter: to collect bits from a liquid passing through

ingredient: a food that is mixed with other foods

physical: having to do with the material world. Physically means able to be done by putting forth effort.

plasma: a substance that is similar to a gas but that can carry electricity

property: special feature

suspend: float

For More INFORMATION

Books

Brown, Cynthia Light. *Kitchen Chemistry: Cool Crystals, Rockin' Reactions, and Magical Mixtures with Hands-On Science Activities.* Norwich, VT: Nomad Press, 2020.

Heinecke, Liz Lee. *Chemistry for Kids: Homemade Science Experiments and Activities Inspired by Awesome Chemists, Past and Present.* Beverly, MA: Quarry Books, 2020.

Heinecke, Liz Lee. *Kitchen Science Lab for Kids: EDIBLE EDITION: 52 Mouth-Watering Recipes and the Everyday Science That Makes Them Taste Amazing.* Beverly, MA: Quarry Books, 2019.

Websites

Chem4Kids
chem4kids.com/files/matter_mixture.html
Learn how a mixture is different from a compound and more!

Kiddle
kids.kiddle.co/Mixture
Kiddle explores the wonderful world of mixtures!

Science Sparks
science-sparks.com/making-mixtures/
Learn how you can make mixtures at home.

Publisher's note to educators and parents: Our editors have carefully reviewed these websites to ensure that they are suitable for students. Many websites change frequently, however, and we cannot guarantee that a site's future contents will continue to meet our high standards of quality and educational value. Be advised that students should be closely supervised whenever they access the internet.

INDEX